NEWSPRINT

CANADIAN SUPPLY AND

AMERICAN DEMAND

THOMAS R. ROACH

The Forest History Society is a nonprofit, educational institution dedicated to the advancement of historical understanding of human interaction with the forest environment. The Society was established in 1946. Interpretations and conclusions in FHS publications are those of the authors; the Society takes responsibility for the selection of topics, the competence of the authors, and their freedom of inquiry.

Grants to the Forest History Society supported work on this book and its publication.

Forest History Society
701 Vickers Avenue
Durham, North Carolina 27701
(919) 682-9319

Library of Congress Cataloging-in-Publication Data

Roach, Thomas R.
    Newsprint : Canadian supply and American demand / Thomas R. Roach.
      p. cm. — (Forest History Society issues series)
    Includes bibliographical references.
    ISBN 0-89030-050-X
    1. Newsprint industry—Canada—History. 2. Newsprint industry—United States—History. I. Title. II. Series.
    HD9839.N43C3667 1994
    338.4'7676286—dc20
                                                              94-6940
                                                                   CIP

Forest History Society Issues Series

The Forest History Society was founded in 1946. Since that time, the Society, through its research, reference, and publication programs, has advanced forest and conservation history scholarship. At the same time, it has translated that scholarship into formats useful for people with policy and management responsibilities. For nearly five decades the Society has worked to demonstrate history's significant utility.

The Forest History Society Issues Series is the latest and most explicit contribution to history's utility. With guidance from the Advisory Commitee, the Society selects issues of importance today that also have significant historical dimensions. Then we invite authors of demonstrated knowledge to examine an issue and synthesize its substantial literature, while keeping the general reader in mind.

The final and most important step is making these authoritative overviews available. Toward that end, each pamphlet is distributed to people with management, policy, or legislative responsibilities who will benefit from a deepened understanding of how a particular issue began and evolved.

The Issues Series—like its Forest History Society sponsor—is nonadvocacy. The series aims to present a balanced rendition of often contentious issues. While all views are aired, the focus is on consensus. The pages that follow document the growth of the Canadian newsprint industry and its traditional reliance on the U.S. market. We see that in more recent years significant market shifts, a strong Canadian dollar, and environmental concerns have brought newsprint's future into question.

The Society gratefully acknowledges financial support from the Canadian Pulp and Paper Association and the Government of Canada for this second title in the Issues Series.

Advisory Committee

William H. Banzhaf, Society of American Foresters
David E. Barron, Canadian Pulp and Paper Association
John L. Blackwell, World Forestry Center
M. Rupert Cutler, Virginia's Explore Park
R. Peter Gillis, Treasury Board of Canada
Digges Morgan, Southern Forest Products Association
Mark A. Reimers, USDA-Forest Service
Roger A. Sedjo, Resources for the Future
Ronald J. Slinn, Slinn and Associates
William D. Ticknor, Forestry Consultants

Contents

Overview vi
Introduction vii

Chapter I: The Early Years 1
   Export Restrictions and Tariffs, 1897-1913 2
   The Effect of Free Trade in Paper 2
   The Canadian Response 4
   The Newsprint Market, 1913-20 6

Chapter II: The Great War and After 9
   Industry Leaders 9
   The Importance of the Market 11
   The Pulpwood Market 11
   1929 and After 12

Chapter III: The Great Depression and the Second World War 14
   The Newsprint Industry 14
   Government Intervention, 1928-39 17
   The End of the 1930s 22
   Industry During the Second World War 22

Chapter IV: The Postwar Years 27
   Pulp and Paper, 1945-50 27
   Newsprint in the 1950s 29

Chapter V: Modern Times 36
   The Environment 36
   Changes in the Structure of the Pulp and Paper Industry 39
   An Economic Overview of Newsprint 41
   Factors Influencing the Price of Newsprint 41
   Market Changes 46
   Canadian Newsprint Manufacturers 48
   Canadian International Competitiveness 49
   Oversupply Not the Problem 51
   Machine Size a Factor 51
   The Changing Structure of Canadian Forests 52
   A New Approach to Fiber Production 53

Suggested Reading 54

Figures

1. United States and Canadian paper production, 1913-20 5
2. Percentage of Canadian newsprint and pulpwood exported to the United States, 1913-20 7
3. Average cost per ton of newsprint and pulpwood, 1913-21 8
4. Relationship between advertising space and newsprint consumption, 1914-20 10
5. Canadian newsprint production and New York price, 1920-30 10
6. Canadian pulpwood production vs. exports, 1920-30 13
7. U.S. advertising, 1929-39 15
8. Canadian production and U.S. consumption, 1929-39 16
9. Newsprint price, f.o.b. New York, 1929-41 21
10. Canadian newsprint production and price, 1939-49 24
11. Canadian newsprint production vs. percentage of capacity, 1939-49 26
12. U.S. newsprint consumption and advertising space, 1945-55 28
13. Canadian newsprint production vs. percentage of capacity, 1945-59 34
14. Newsprint manufactured in the United States, 1945-59 35
15. U.S. newsprint consumption, 1960-90 40
16. Canadian newsprint production vs. percentage of capacity, 1960-90 42
17. Newsprint price, f.o.b. New York, 1960-90 43
18. U.S. imports of Canadian newsprint, 1960-90 47
19. Value of Canadian dollar in terms of U.S. dollars, 1952-89 50

Tables

1. Newsprint mills in Canada, 1920 3
2a. Canadian-owned newsprint companies, 1934 18
2b. American-owned newsprint companies in Canada, 1934 19
3. International Paper Co. wood conversion plants in Canada, 1934 23
4a. Canadian-owned newsprint companies, 1948 30
4b. American-owned newsprint companies in Canada, 1948 31
5. New capacity in the southern United States, 1953-59 32
6. Canadian wood products as percentage of total, 1948, 1960, 1985 38
7. Canadian newsprint manufacturers, June 1991 44

Overview

Although newsprint is still the largest single product from Canada's pulp and paper industry, it has decreased in importance over the past thirty years. Market shifts, currency exchange rates, and environmental concerns are important parts of this story:

- Since the end of the Second World War, the scope of Canada's pulp and paper industry has broadened and diversified.

- Canada's paper manufacturers are successfully seeking and capturing product niches for themselves.

- Environmental controversy, which has followed Canada's forest industry since the beginning of the conservation movement in the late-nineteenth century, still continues.

- U.S. requirements that newsprint be manufactured from at least 50 percent recycled material affects Canadian industry.

- Since the 1960s Canadian newsprint manufacturers have maintained production at between 85 percent and 100 percent of capacity.

- In 1960 Canada manufactured 70 percent of newsprint consumed in the U.S.; currently this figure is close to 50 percent.

- Exchange rates, not production costs, have been the major factor in determining Canada's competitive position.

Introduction

The vitality and size of the Canadian pulp and paper industry is largely the result of American newsprint demand, and newspaper consumption drives demand for newsprint. It is this demand that has historically set Canadian prices and priorities. The modern newspaper traces its origins to peculiar circumstances in the United States during the final years of the First World War, when advertisers combined the effectiveness of mass advertising with newspapers as a medium for delivery. This combination led to a tremendous increase in newspaper consumption in North America. Canada's newsprint industry grew rapidly during the years between the two World Wars, but suffered later from overexpansion, loose investment practices, and misunderstanding of the dynamics between mass advertising and the economic health of the United States and Canada.

Canada gained economically during the Second World War. The newsprint industry, however, changed little from 1939-49. During the 1950s Canada's newsprint manufacturers prospered, expanding operations and modernizing machines but building no new mills. Failure to build during the prosperous 1950s contributed to the industry's current economic difficulties.

Canada has recently faced considerable competition from new mills both in the southern United States and overseas. Investment in modern mills and drastic changes in forest management practices could help Canadian newsprint manufacturers regain their former leadership position. Canada can have an economically sound and environmentally sustainable newsprint industry, but the future will tell whether the industry in Canada accepts this challenge.

The Early Years

The first rag-based paper mill in Canada was constructed at St. Andrews, Quebec, in 1804-1805. Three weekly Montreal newspapers were the mill's apparent market. Groundwood pulp, first used in 1867 to produce paper in the United States, started being used in Canada shortly thereafter. The Riordon Company of Ontario pioneered Canadian use of groundwood pulp to produce newsprint, while the Toronto Paper Company first introduced newsprint production using sulfite pulp in 1886.

In North America, use of groundwood paper grew slowly at first, but in the last two decades of the nineteenth century annual consumption increased from 452,000 tons to over 2,000,000 tons. Several factors account for this phenomenal expansion. Both Canada and the United States enjoyed steady population growth, while from 1880-1900 the cost of paper fell as production was mechanized and the number of mills increased. At the same time, newspaper readership increased dramatically in both countries.

In the United States paper production from wood pulp faced an important restriction. The mechanical and chemical methods used to convert trees to paper during this period worked best with a limited selection of tree species. White and black spruce were the only species suitable, and in 1910 spruce made up 58 percent of the wood used in the United States to make newsprint. By the 1930s the choice widened to include most pines and members of the poplar family. Spruce remained the tree of choice but there was limited supply of northern spruce in the United States compared with the supply in Canada.

Demand for northern spruce soon outstripped supply as newspaper circulation in the U.S. grew rapidly during the 1880s and 1890s. When mills found themselves running out of wood, the logical solution was to import pulpwood from Canada since at the time presses were equipped to use the type and grade of paper that northern spruce produced.

Export Restrictions and Tariffs, 1897–1913

A major obstacle to U.S. imports was that provincial governments controlled most Canadian pulpwood forests. To protect Canadian conversion companies, there were restrictions on exporting logs cut from Crown lands. By 1900 Canadian federal and provincial governments promulgated restrictions on exporting wood cut from lands they controlled. Typically, these restrictions required that the wood be partially processed or "manufactured" before being exported. The degree of processing varied among jurisdictions but all prevented exporting raw logs cut from Crown lands to processing plants in the United States.

To retaliate for these restrictions, in 1897 the United States imposed a tariff on paper products imported from Canada. This tariff was high enough to price Canadian newsprint out of the American market. The effect in Canada was to halt the building of several mills in Ontario. By 1905 purchases of partially processed wood or pulpwood from private landowners in Canada could not offset the growing shortage of pulp logs in the United States. To the anger of American newspaper publishers, the price of newsprint began to rise. As this happened, publishers lobbied vigorously to remove the tariffs on Canadian paper. They succeeded despite counterpressure from American pulp mills that profited from higher prices. In 1911 the U.S. government lowered the tariff on imported paper and in 1913 completely removed it.

The Effect of Free Trade in Paper

The 1913 removal of the imported paper tariff had both immediate and long-lasting effects on the Canadian pulp and paper industry. In the short time before World War I began, several American companies built mills in Canada and Canadian companies started either building new plants or expanding existing plants.

The Powell River Company, controlled by the Brooks Scanlon Lumber Company of Minneapolis, and Pacific Mills, a subsidiary of the Crown Willamette Paper Company of San Francisco, were two important American companies that built plants in Canada at this time. These two mills were the first wood-based paper plants established in British Columbia. The Chicago Tribune created the Ontario Paper Company and constructed a mill at

Table 1 Newsprint mills in Canada, 1920

Mill	Location	Capacity (tons)
Abitibi Power & Paper Co.	Iroquois Falls, Ontario	237
Belgo-Canadian P&P Co.	Shawinigan Falls, Quebec	206
J. R. Booth	Ottawa, Ontario	141
Brompton P&P Co.	East Angus, Quebec	91
Canada Paper Co.	Windsor Mills, River St. Francis, Ontario	36
Donnacona Paper Co.	Donnacona, Quebec	106
E. B. Eddy Co.	Hull, Quebec	54
Fort Frances P&P Co.	Fort Frances, Ontario	125
Laurentide Co.	Grand-Mère, Quebec	224
News Pulp & Paper Co.	St. Raymond, Quebec	32
Ontario Paper Co.	Thorold, Ontario	218
Pacific Mills	Ocean Falls, British Columbia	216
P. C. Powell River Co.	Powell River, British Columbia	228
Price Brothers & Co.	Jonquière & Kenogami, Quebec	260
Riordon Bros.	Merritton & Hawkesbury, Ontario	224
Spanish River P&P Mills	Espanola, Sault Ste. Marie, & Sturgeon Falls, Ontario	489
St. Maurice Paper Co.	Trois Rivières, Quebec	111

Source: Royal S. Kellogg, *Newsprint Paper in North America* (New York: The Newsprint Service Bureau, 1948), p. 24.

Thorold, Ontario. Whereas the two mills in British Columbia operated with wood cut on their own limits, the Ontario Paper Company intended to rely on wood cut by farmers and settlers.

The Canadian Response

The largest paper manufacturer in Canada grew out of a small regional paper mill founded by the Riordon family in 1857. By the 1870s this mill had expanded, and using increasing quantities of wood pulp it became the sole supplier of newsprint to the Toronto Globe. During the 1880s and 1890s, the Riordon mills were the only newsprint manufacturers in Ontario, a position they carefully maintained by adjusting prices downward in step with American mills. By the 1890s papermaking in Ontario had broadened. The Toronto Paper Company in 1886 opened Ontario's first sulfite mill, producing a finer quality paper than the groundwood-based newsprint. Riordon's excursion into sulfite in 1890 was a success, and the company opened its first all-sulfite mill at Hawkesbury in 1898.

The Riordon Company was positioned to take advantage of the opportunities created by the events of 1911-13. Removal of the tariff also contributed to the fortunes of the Laurentide Paper Company of Grand-Mère, Quebec. Laurentide later developed innovative and effective forest management programs. A group that gained control of a paper mill at Sturgeon Falls in northern Ontario also exploited the opportunities provided by the free access of newsprint to the United States market. From 1911-13 control of this plant moved from one company to another, finally becoming part of the Spanish River Pulp and Paper Company.

Several other companies began producing wood pulp for export. These included two Ottawa Valley mills owned by the E. B. Eddy Company and the J. R. Booth family, and the Belgo-Canadian Company. Belgo-Canadian, founded at the turn of the century, produced wood pulp for export to Belgium. Other mills also opened in the years between 1913 and 1920. Most notable was the Dryden Paper Kraft paper mill that began production in 1918. This mill produced a heavy brown colored paper suitable for packaging.

Figure 1 **United States and Canadian paper production, 1913–20**

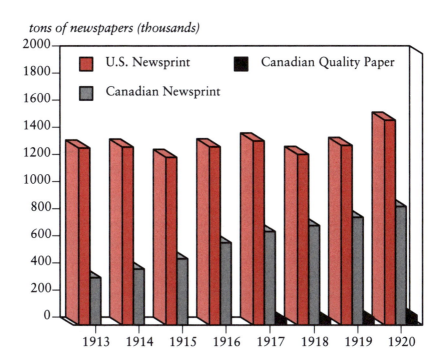

By the end of the First World War, papermaking in Canada was set on a new course. Pulp and paper production steadily increased over the years, and by 1920 one hundred mills produced pulp and a wide range of papers in Canada; seventeen major mills in Canada and Newfoundland produced newsprint. Meanwhile U.S. production remained stable.

The Newsprint Market, 1913–20

By the end of the First World War, papermaking in Canada was on a new course. Pulp and paper production steadily increased over the years, and by 1920 one hundred mills produced pulp and a wide range of papers in Canada; seventeen major mills in Canada and Newfoundland produced newsprint.

United States production grew only slowly from 1913-20 while Canadian production more than doubled (see figure 1). In fact most Canadian production was exported to the United States, and in 1925 the United States imported more newsprint from Canada than it manufactured for itself. Canadian newsprint exports to the United States jumped from approximately 42 percent of total production in 1913 to over 77 percent in 1920 (see figure 2). At the same time, pulpwood production from private lands in Canada rose while the amount exported to the United States dropped from 64 percent of the total cut in 1908 to 31 percent in 1920.

The economic and market events that occurred in the years 1913-20 had a tremendous effect not only on the mills built prior to this period but on mills constructed during the 1920s. A trend started in Canada's papermaking industry: although newsprint production increased greatly in this period, production of other papers grew only slowly (see figure 1). Soon the Canadian papermaking industry became almost exclusively a manufacturer of newsprint. By the late 1920s, over 90 percent of the newsprint Canada produced was sent to the United States.

In 1913 newsprint cost about $45 per ton delivered to New York (see figure 3). For the next two years this average price fell slightly, largely because of increasing imports from Canada. The situation reversed, however, as wartime controls on production by the Canadian government, labor shortages, increasing demand from American purchasers, and United States entry into the war all combined to shorten the available paper supply. The delivered price of newsprint to New York peaked in 1920 at $112.60 per ton, a price it would not reach again until the 1950s.

Steady growth in product demand, coupled with large price increases and low costs for pulpwood, had a heady effect on paper manufacturers. The problem was that this boom for Canadian newsprint encouraged overinvestment in new capacity. The market and its dynamics were still new and the phenomenal success of advertising campaigns carried out at the end of the war was so intense and attractive that businessmen were misled. These mistakes in overinvestment and expansion haunted the industry for twenty years, until the outbreak of the Second World War.

Figure 2   **Percentage of Canadian newsprint and pulpwood exported to the United States, 1913–20**

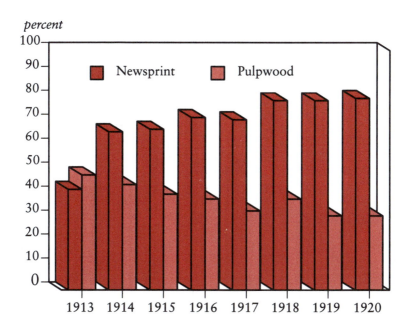

Most Canadian newsprint was exported to the United States. This trend continued, and by 1925 the United States imported more newsprint from Canada than it manufactured for itself. Pulpwood exports, however, declined as the domestic market absorbed surplus production.

Figure 3 Average cost per ton of newsprint and pulpwood, 1913–21

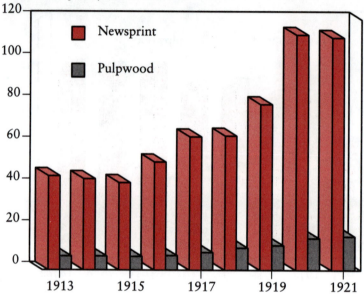

Wartime controls on production by the Canadian government, labor shortages, increasing demand from American purchasers, and United States entry into the First World War all combined to shorten the available paper supply. The delivered price of newsprint to New York peaked in 1920.

The Great War and After

Demand for advertising space, premised on the example of the successful Liberty Bond advertising campaigns of 1917 and 1918, expanded rapidly once peace was declared. Contemporary accounts indicate that after the First World War demand for advertising was so great that many United States newspapers had to limit advertising space. This demand encouraged newsprint producers to invest in new mills on the expectation that newsprint consumption would continue to grow through the 1920s.

The bases of this expansion were two phenomena experienced in 1919-21: a steep rise in the price of newsprint and the expectation that newspaper advertising would continue to expand (see figure 4). But these expectations were based on false premises. In fact the cost of advertising limited its effectiveness, which slowed the expansion in newsprint demand. Because the newsprint market was limited, supply soon outgrew demand and the market behaved predictably: prices fell. From a high of $112.60 per ton delivered to New York in 1920, the price fell to $62.00 per ton in 1930 and then tumbled further to a low of $40.00 per ton in 1934-35. Production was at 99 percent of capacity in 1920 but only at 54 percent of capacity in 1933, in part due to the Great Depression.

Industry Leaders

Once import tariffs were removed in 1913, International Paper (IP) began producing fine papers instead of newsprint at its American mills, since fine papers were still protected by tariffs. At the same time IP began to concentrate newsprint production at its Canadian mills, a process completed in 1930. To

Figure 4 **Relationship between advertising space and newsprint consumption, 1914–20**

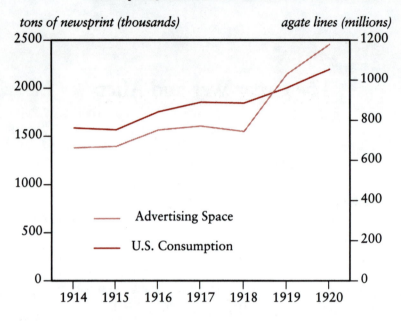

Figure 5 **Canadian newsprint production and New York price, 1920–30**

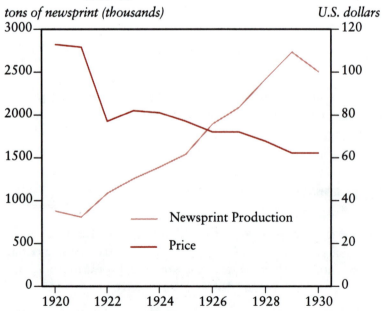

Source for figure 5: Royal S. Kellogg, *Newsprint Paper in North America* (New York: The Newsprint Service Bureau, 1948).

this end, the company constructed a new mill at Trois Rivières, Quebec, that opened in 1921. Shortly thereafter, IP accelerated expansion of newsprint production in Canada by acquiring the Riordon family holdings. Using these holdings IP founded a wholly-owned Canadian subsidiary, Canadian International Paper (CIP). By 1930 CIP's daily newsprint production capacity was 1,817 tons; the company owned a 49 percent interest in the E. B. Eddy Company and had a controlling stake in the Bathurst Power and Paper Company of New Brunswick. But most important, CIP had contracts with the Hearst newspaper chain in the United States for the purchase of almost all the paper the company could produce.

The Importance of the Market

From 1921 to 1935 Canadian newsprint production rose while the price fell (see figure 5). CIP could usually underprice its competition because the company had contracts that enabled its mills to run at close to 90 percent of capacity. CIP mills were also newer and more efficient than those owned by other Canadian companies. The mills of other companies in Canada were forced to operate at lower levels of production, sometimes approaching 40 percent of capacity. The market leader had thus placed the problems caused by excess capacity on the competition.

The Pulpwood Market

A lively trade in pulpwood cut from privately owned lands, especially in Ontario, Quebec, and the Maritimes, emerged during the years between the two World Wars. A network of dealers handled this trade. Each dealer covered a specific area of the country, and by mutual consent there was little overlapping. Often the dealer was a local merchant who would advance money in the fall in return for a specified quantity of pulpwood cut by the farmer and his family during the winter.

On its face the pulpwood trade appeared to assist farmers and settlers by providing them with cash income during the winter. The governments of Ontario and Quebec were particularly keen to see the northern Clay Belt area settled with farmers. To assist in the settlement process these governments promoted the idea that harvesting pulpwood while clearing land was the farmer's "first crop," which would effectively cover the costs of establishing a farm. In reality money received for pulpwood cut in the winter did not cover the real cost of clearing land. This disparity resulted in considerable social stress as well as high rates of land abandonment by new settlers.

The situation of the settlers worsened through the latter half of the 1920s and during the Great Depression. Although private pulpwood production remained fairly steady, the value of the product fell as the price of newsprint and the quantity of pulpwood exported to the United States steadily decreased. This trend was disastrous from pulpwood producers' viewpoint because wood exported to the United States had paid one or two dollars more per cord than wood sold to Canadian mills (see figure 6).

Canadian newsprint producers, however, were happy with this situation. By the mid-1920s most Canadian mills relied on a timber limit for some pulpwood. All used contractors to cut the wood and get it to the mill. Because privately cut wood was sold on the oversupplied open market, its cost was lower than the price demanded by contractors for cutting on the mills' limits. Depending on price and convenience, a mill in Ontario might purchase from privately owned sources 30 percent to 60 percent of the wood it needed. Mill owners therefore used the threat of purchasing privately cut wood to reduce demands made by contractors. Also, mills could buy privately cut wood on an "as needed" basis, allowing greater flexibility in purchases of raw material.

1929 and After

The financial markets' collapse in October 1929 ended the boom years. As the Great Depression began, Canadian-owned paper producers were in a bad position because they had all employed interest-paying stocks to finance expansion and takeovers. The stock market crash left companies with considerably overvalued assets. Furthermore, the demand for newspaper advertising space responded negatively to problems in the economy. Soon after the market plummeted, demand for advertising space fell and newsprint consumption fell along with it. Most Canadian newsprint producers therefore experienced financial difficulties during the 1930s. They responded with two tactics. The first was to revalue their preferred stocks and bonds so as to lower annual interest charges. The second was a series of attempts to control newsprint production. For this second task companies looked to the provincial and federal governments of Canada for assistance.

Figure 6   Canadian pulpwood production vs. exports, 1920–30

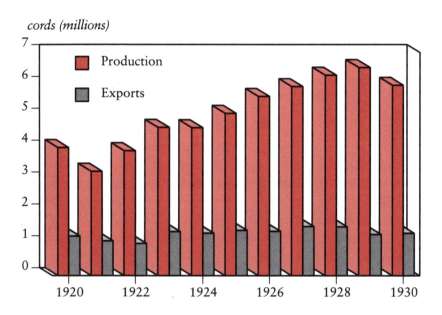

As Canadian newsprint production increased during the 1920s, the quantity of pulpwood Canada exported to the United States remained steady or decreased. This situation was disastrous from the viewpoint of pulpwood producers because wood exported to the United States paid one or two dollars more per cord than wood sold to Canadian mills.

The Great Depression and the Second World War

The Great Depression created a financial environment in North America that was in complete contrast to the 1920s. America's gross national product fell from $87.4 billion in 1929 to $41.7 billion in 1932. The number of unemployed, four million in 1930, rose to eight million in 1931 and reached twelve million by 1932. The Federal Reserve's index of manufacturing production, at 110 in 1929, dropped to 57 by 1932. Total wages paid to Americans declined from $50 billion to $30 billion.

During the 1920s and 1930s, Canada exported 80 percent of its production from forests, mines, and farms. The country's economic stability depended on markets over which it had no control. By 1932 Canadian wheat fetched one-third the price it commanded immediately before the crash, and business investment in new equipment was one-fifth its 1929 level. Economic improvements in the United States during the late 1930s and early 1940s finally helped the recovery of the Canadian newsprint industry.

The Newsprint Industry

Canada exported about 90 percent of newsprint production to the United States at the beginning of the Depression. However the poor financial situation quickly affected newsprint consumption because it meant as much as a one-third reduction in purchased advertising space (see figure 7). People often could not afford advertised products anyway, and many no longer even bought newspapers regularly. By mid-decade and until the beginning of the Second World War sales recovered slowly. Canadian newsprint production and per capita consumption both reflect this American trend during these years (see figure 8).

Figure 7   U.S. advertising, 1929–39

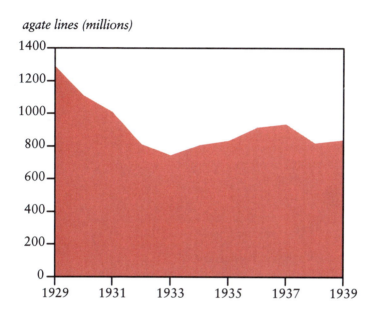

The poor financial situation caused by the Great Depression quickly affected newsprint consumption because it meant a reduction in advertising space purchased. People often could not purchase advertised products anyway, and many people no longer even bought newspapers regularly. An agate line is a unit of measurement for classified advertising.

Figure 8   **Canadian production and U.S. consumption, 1929–39**

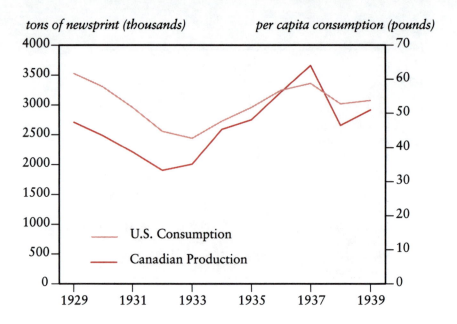

Sources: Royal S. Kellogg, *Newsprint Paper in North America* (New York: The Newsprint Service Bureau, 1948) p. 31; John A. Guthrie, *The Newsprint Paper Industry* (Cambridge, Massachusetts: Harvard University Press, 1941), p. 234.

Canadian newsprint production and U.S. consumption decreased in tandem at the beginning of the Great Depression. Sales and consumption recovered slowly until the beginning of the Second World War.

Canadian newsprint producers found themselves in trouble. The first large company to default on its bonds was the Canada Power and Paper Company (CPPC). In the late 1920s its chairman, Sir Herbert Holt, tried to purchase as many small "independent" newsprint producing companies as possible. He wanted to create the largest Canadian-owned newsprint company and thereby wrest control of the market from IP. Having obtained the assets of companies with a combined value of $60 million, CPPC issued debentures worth $160 million to support more acquisitions. The plan fell through with the 1929 market crash and the subsequent Great Depression. CPPC eventually reorganized as the Consolidated Paper Company in 1931. By the end of 1932, Abitibi Power and Paper and the independent St. Lawrence Corporation had also defaulted; companies controlling 58 percent of total newsprint production capacity in Canada had either passed into receivership or voluntarily reorganized (see table 2).

Government Intervention, 1928-39

Government intervention in the Canadian pulp and paper industry dates back to the First World War. The government eventually lifted price and production controls early in 1919. During the first half of the 1920s Canadian federal and provincial governments looked with favor on the industry's expansion and investment in new plants. By 1926-27, however, this attitude changed when it became obvious that the industry's overexpansion coupled with reductions in the price of newsprint would lead some companies into financial difficulty.

Two reasons account for this change in attitude toward the industry and for subsequent government involvement. The first was that all mills used access to a public resource to provide them with raw material. With the price of newsprint falling, and the possibility that some mills would go out of business, governments grew concerned that the resource would be overexploited as companies high-graded their stands in an attempt to reduce costs further. Implicit in this publicly stated argument was concern that the resource would be sold at prices below real value. The second reason, never publicly stated but much more persuasive, was that the industry provided a livelihood for thousands of mill workers. This was particularly true in Ontario and Quebec where the mills increasingly located in remote northern areas of the provinces. In addition, each company employed hundreds of men and horses in the woods cutting pulpwood. As the 1930s passed, these men moved

Table 2a   Canadian-owned newsprint companies, 1934

Name of company	Daily capacity (in tons)
Abitibi Power & Paper Co. (six mills, G. H. Mead owned a large block of shares)	2,013
Anglo-Canadian P&P Mills, Ltd. (part of Consolidated Paper Corp.)	480
Bathurst Power & Paper Co. (owned by Newsprint Bond & Share Corp., operated by International Paper Co.)	140
J. R. Booth	158
Brompton P&P Co. (formerly part of St. Lawrence Paper Mills, Ltd.)	239
Consolidated Paper Co. (six mills)	1,944
E. B. Eddy Co. (49 percent of stock owned by International Paper Co.)	127
Lake St. John Power & Paper Co. (formerly part of St. Lawrence Paper Mills, Ltd.)	260
Maclaren Co.	240
Mersey Paper Co.	293
News Pulp & Paper Co.	37
Price Brothers & Co. (one mill in receivership)	1,020
St. Lawrence Paper Mills, Ltd.	460
Thunder Bay Paper Co. (controlled by Abitibi Power & Paper Co.)	246

Table 2b — American-owned newsprint companies in Canada, 1934

Name of company	Daily capacity (in tons)
Beaver Wood Fibre Co.	77
Canadian International Paper Co. (see table 3, p. 23)	1,817
Donnacona Paper Co. (previously owned by Price Brothers & Co.)	240
Fort Frances P&P Co. (part of the Backus Group)	238
Great Lakes Paper Co. (part of the Backus Group)	314
Kenora Paper Mills, Ltd. (part of the Backus Group)	252
Ontario Paper Co. (owned by the Chicago Tribune)	429
Pacific Mills, Ltd.	256
P. C. Powell River Co.	650
Spruce Falls Power & Paper Co.	480

Sources for tables 2a and 2b: Herbert Marshall, Frank Southard Jr., and Kenneth W. Taylor, *Canadian-American Industry: A Study in International Investment* (Toronto, Ontario: McClelland and Stewart, 1976), Appendix III.

from being seasonal employees of contractors to year-round employees of the mills. This trend was especially noticeable in Ontario. Maintaining year-round operation of the northern mills became a political imperative for the provincial governments.

In 1927 the leading Canadian newsprint producers and their selling organizations formed the Canadian Newsprint Company in an attempt to control prices and prevent price cutting. The organization lasted less than a year. It collapsed when two member companies withdrew in a quarrel over pricing policies. Almost immediately, the Newsprint Institute of Canada was formed with the same objectives as its predecessor. The institute, however, enjoyed considerable support from the governments of Ontario and especially Quebec. Even so, it ceased to function early in 1931. After this date attempts to control the market were made by appealing directly to provincial political leaders.

Both the Canadian Newsprint Company and the Newsprint Institute of Canada collapsed because of the large quantity of newsprint that was available. Newsprint was flooding the market, keeping prices low (see figure 9). Both organizations attempted to persuade member companies to cut back on production and hold to a common price. The arguments failed either to convince the market leader (usually IP) or to benefit some member companies, whose financial positions were worsening.

Once a company fell into the hands of its bondholders or receivers, production restraints disappeared. By 1932 bankrupt companies were dumping paper onto the spot market as well as giving under-the-table discounts on contract prices that were already only a few cents per ton above production costs. This was because so many new plants were built in the 1920s that investors could not get their money back by selling the plants. The only way creditors and bondholders could hope to salvage their investment was to keep the mills operating.

The industry increasingly appealed for government intervention during the 1930s. In 1929-30, for example, Premier Louis Alexandre Taschereau requested that a price of $60 per ton be maintained for newsprint; he summoned IP President Archibauld Graustein to insist that the company set and maintain this price. This demand was to no avail since the American Newspaper Publishers' Association responded with threats to purchase from equally hard-pressed European producers.

After the 1932 election of President Franklin D. Roosevelt in the U.S., the initiative for government intervention through its National Recovery Administration passed to the United States. The NRA success at controlling newsprint price is debatable; even the official history of the NRA concedes

Figure 9 **Newsprint price, f.o.b. New York, 1929–41**

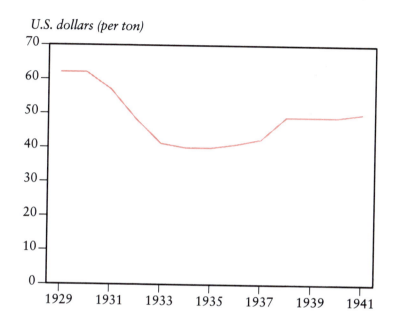

Source: Royal S. Kellogg, *Newsprint Paper in North America* (New York: The Newsprint Service Bureau, 1948) p. 50.

Newsprint flooded the market during the 1930s, keeping prices low. Protectionist business groups collapsed because of the large quantity of newsprint that was available. These organizations had attempted to persuade member companies to cut back on production and hold to a common price.

that improvements in the industry resulted from a general improvement in business conditions. The provincial governments of Ontario and Quebec, however, supported efforts of the American agency and backed their support with threats to increase Crown dues on wood cut by companies that failed to maintain agreed-upon prices. The provinces also pressed demands that Canada's federal government prevent the export of cheap paper by these companies.

Quebec took other actions as well. In 1936 the Ontario Paper Company decided to build a new mill in a remote area of the province. Although its paper was not sold on the open market, Ontario Paper's parent company, the Chicago Tribune, had contracts with other mills and purchased newsprint on the spot market. The subsidiary's planned expansion would clearly lower these purchases. The Canadian industry lobbied the province's new premier, Maurice Duplessis, to withhold permission to build the new mill. These efforts failed because the mill was being built in a region with high unemployment. Duplessis, however, required that the mill be owned by a company registered in Quebec and that the license agreement be renegotiated. These requirements cost the Chicago Tribune considerable money, but the expansion continued.

The End of the 1930s

By the end of the decade improvements in the U.S. economy led Canada's newsprint industry to anticipate better times. Although there had been considerable reorganization within the industry, the massive mergers that had been predicted never took place. As a result of problems businesses encountered in this period, American capital owned 40 percent of the productive capacity of newsprint in Canada by the middle of the decade. Canadian companies, however, produced 50 percent of the newsprint actually sold in North America. Much of this was produced by IP, which by 1934 was the second largest newsprint producer in Canada (see table 3).

Industry During the Second World War

Canada declared war on Germany 10 September 1939, but effects on the newsprint industry did not appear until the following spring. By then the Canadian and British governments had requisitioned all suitable ships for use

Table 3   **International Paper Co. wood conversion plants in Canada, 1934**

Company	Product and location
Continental Paper Products	Paper bag mill: Ottawa, Ontario
Continental Wood Products	Lumber mill: Elsas, Ontario
International Fibre Board, Ltd.	Fiberboard mills: Gatineau, Quebec; Midland, Ontario

Canadian International Paper Co.	Paper mills:
	  Gatineau & Trois Rivières, Quebec;
	  Dalhousie, New Brunswick
	Pulp mills:
	  Timiskaming, Quebec; Hawkesbury &
	  Nipigon, Ontario
	Pulpwood preparation plants:
	  Gaspe, Pentecost, Hull, Batiscan, &
	  Trois Rivières, Quebec; Rockland, Ontario
	Lumber mills:
	  Calumet & Cap de la Madeleine, Quebec
	Coal mine:
	  Minto, New Brunswick

Newfoundland Power & Paper	Pulp & paper mill: Corner Brook, Newfoundland

Source: Herbert Marshall, Frank Southard Jr., and Kenneth W. Taylor, *Canadian-American Industry: A Study in International Investment* (Toronto, Ontario: McClelland and Stewart, 1976), p. 41.

---

on the Atlantic to carry supplies to Britain and to British forces fighting in France. This action affected some companies' ability to transport pulpwood and newsprint.

Newsprint price continued to be set by market demand, and from 1939 until the United States entered the war late in 1941, it remained at $50 per ton f.o.b. New York (see figure 10). At this point production and price controls were imposed on the industry from both sides of the border. In effect, the United States Office of Price Administration (OPA) set the price of newsprint

Figure 10 Canadian newsprint production and price, 1939–49

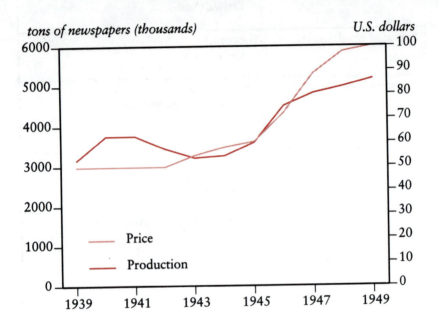

Because of the war there were controls on the price of newsprint and the quantity of newsprint that could be produced. In 1941 the U.S. Office of Price Administration began setting prices on the American market, where Canada shipped 90 percent of its exports. Starting in January 1943, the United States War Production Board ordered all newspaper publishers to limit newsprint consumption to 1941 levels, plus 3 percent for waste. By 1945 newsprint was undervalued.

on the American market. Because Canada exported over 90 percent of its production to the U.S., this meant the OPA set the price for Canada as well. Although prices increased to reflect inflation, when the war ended in 1945 the price of newsprint was considerably below its real value.

There were also controls on the quantity of newsprint that could be produced. Starting in January 1943, the United States War Production Board ordered all newspaper publishers to limit newsprint consumption to 1941 levels, plus 3 percent for waste. Thus American publishers had to reduce the size of their newspapers and in the process decide how to balance news against advertising space. Many papers solved this problem by selling advertising based on the edition in which it would appear. Some papers restricted the size of advertisements. Further cuts in consumption followed in late 1943 and in 1944. The effects of these reductions are shown in figures 10 and 12. In Canada not only did newsprint production drop, but so did the use of manufacturing equipment as a percentage of full production capacity (see figure 11).

The Canadian Wartime Prices and Trade Board created a system that controlled production at all paper mills. The board had power to penalize producers who exceeded their quotas as well as to compensate those whose production fell short. This was done through a pool arrangement, with penalties paid into the pool and compensation paid from it.

Critics of Canadian production control regulations argued that the pool arrangement had several adverse effects. They believed it tended to reward companies located in isolated parts of the country that did not compete with other industries for electrical power because they generated it themselves. Furthermore, remote companies had "captive" labor that was unavailable for work in plants producing war material without relocating mills. Older, inefficient mills with large capacities often received higher allocations than newer mills, even though the production cost per ton was lower for newer mills.

Generally the regulations ensured that the Canadian newsprint industry survived the war relatively unchanged. Most mills operating at the end of the 1930s were still operating in 1948. However, labor and manufacturing shortages during the war combined with a low paper price and production quotas to ensure that plants were not modernized. In short, at the end of the Second World War Canada's newsprint industry was outdated, inefficient, and uncompetitive. Luckily, both price and demand for newsprint rose in 1946 and 1947, while the industry had a monopoly. By 1949 Canada's newsprint mills produced paper at 102 percent of its stated capacity, and the price peak of $100 per ton had not been seen since the end of World War I.

Figure 11 **Canadian newsprint production vs. percentage of capacity, 1939–49**

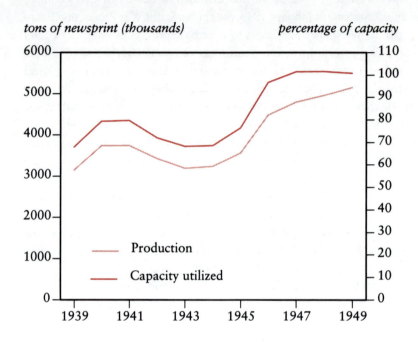

Not only did Canadian newsprint production drop during the Second World War but so did the use of manufacturing equipment (shown as a percentage of production capacity).

The Postwar Years

At the end of the Second World War, Canada's newsprint industry had both a potential for disaster and an opportunity for growth because the war had been particularly hard on European pulp and paper manufacturers. Before the war these companies had provided the only real competition to Canadian dominance of the North American market. Canadians interpreted European difficulties in the immediate postwar years as an opportunity for Canadian manufacturers to tighten their hold on the North American market and to expand throughout the world. By 1946 Canada exported more goods to the United States than it did to the United Kingdom, reversing the prewar situation. Canada's GNP grew from about $12.5 billion to almost $22 billion by 1955, a growth of 76 percent. Much of the expansion, however, was in new secondary manufacturing industries rather than in traditional industries like pulp and paper.

The Korean War dominated the economy during the early 1950s. It produced increased demand for raw materials and government expenditures to support Canada's military efforts. These expenditures later reduced the available capital for funding expansion in the pulp and paper industry as heavy industry became preoccupied with supplying war material. At the same time the Korean War only marginally increased the demand for paper.

Pulp and Paper, 1945–50

Controls implemented during the Second World War, along with the diversion of building supplies and equipment manufacturing facilities to the war effort, left the industry with facilities from the 1930s once the war ended.

Figure 12 U.S. newsprint consumption and advertising space, 1945–55

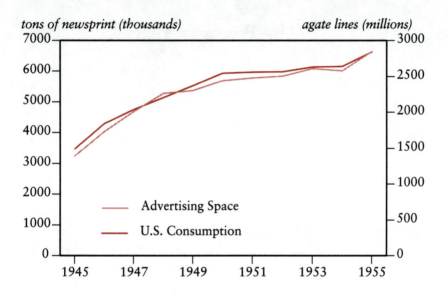

The years 1945-55 were good for Canadian pulp and paper. Demand for newsprint in the United States grew as purchases of newspaper advertising increased. By 1950 the United States used six million tons of newsprint annually and by 1955 nearly seven million tons, a considerable increase from the three and a half million tons used at the end of the Second World War.

The Postwar Years   29

Industry organization and age, quality, and efficiency of equipment were all Depression vintage. Significant numbers of new paper machines were not installed until the 1950s, when the industry finally began to recover from the trauma of the 1930s.

The years 1945-50 were good for Canadian pulp and paper. Demand for newsprint in the United States grew as purchases of newspaper advertising increased (see figure 12). By 1950 the United States used six million tons of newsprint annually, a considerable increase from the three and a half million tons used at the end of the Second World War. Canadian mills continued to hold a near monopoly on the North American market. As it had for years, Canada supplied the United States with over 80 percent of its newsprint requirements. At the same time, Canadian paper manufacturers declared shareholder dividends that doubled from 2.5 percent of operating revenues in 1945 to 5 percent in 1948. According to the Newsprint Association of Canada, these were the first dividends most companies had paid in twenty years or more.

Steady growth lasted from 1945-56. During this time Canadian mills maintained steady production levels near 100 percent of industry capacity (see figure 13).

Newsprint in the 1950s

Trends that were observable in the 1950s became dominant market forces in the Canadian newsprint industry of the 1970s and 1980s. By 1950, for example, it was clear that Canada would not regain the position it held before the war as newsprint supplier to Europe. This market was lost largely because of the high value of the Canadian dollar (see figure 19, p. 50). Throughout the 1950s the Canadian dollar maintained a value close to that of its American counterpart; sometimes it was priced at a premium. High currency values did not help Canada's exports, but did help American investors purchase large blocks of Canada's secondary manufacturing industry. In 1945, for example, the U.S. long-term capital investment in Canada was almost $5 billion and direct investment was $2.3 billion. By 1955, capital investment stood at $10.3 billion and annual direct investment was $6.5 billion. This represented increases during the decade of 106 percent and 182 percent respectively.

Expansion of newsprint production in Canada up to 1950 resulted mostly from activating idle capacity. The first new paper machine to make newsprint in Canada did not start operation until 1949. Although there were plans for

Table 4a   Canadian-owned newsprint companies, 1948

Owner	Mill location
Abitibi Power & Paper Co. (see also Manitoba, Thunder Bay, and Ste. Anne paper companies)	Thunder Bay, Iroquois Falls, Sault Ste. Marie, & Sturgeon Falls, Ontario
Anglo-Canadian P&P Mills, Ltd.	Quebec, Quebec
Brompton P&P Co.	Bromptonville, Quebec
Consolidated Paper Co.	Grande-Mère, Port Alfred, Shawinigan Falls, Trois Rivières, & Cap de la Madeleine, Quebec
E. B. Eddy Co. (49 percent of stock owned by International Paper Co.)	Hull, Quebec
Lake St. John Power & Paper Co. (parent of St. Lawrence Paper Mills, Ltd.)	Dolbeau, Quebec
Manitoba Paper Co. (Abitibi Power & Paper Co. majority stockholder)	Pine Falls, Manitoba
The James Maclaren Co.	Buckingham, Quebec
Mersey Paper Co.	Brooklyn, Nova Scotia
Price Brothers & Co. (reorganized in 1937)	Kenogami, Quebec
Provincial Paper, Ltd.	Thunder Bay, Ontario
Ste. Anne Paper Co. (controlled by Abitibi Power & Paper Co.)	Ste. Anne de Beaupré, Quebec
St. Lawrence Paper Mills, Ltd.	Trois Rivières, Quebec
Thunder Bay Paper Co. (controlled by Abitibi Power & Paper Co.)	Thunder Bay, Ontario

Table 4b American-owned newsprint companies in Canada, 1948

Owner	Mill location
Beaver Wood Fibre Co.	Thorold, Ontario
International Paper Co.	Gatineau & Trois Rivières, Quebec; Dalhousie, New Brunswick
Donnacona Paper Co. (owned by Howard Smith Paper Co.)	Donnacona, Quebec
The Ontario-Minnesota P&P Co. (formerly Fort Frances P&P Co.)	Fort Frances, Ontario
Great Lakes Paper Co.	Thunder Bay, Ontario
The Kalamazoo Vegetable Parchment Co. (mill purchased from Abitibi Power & Paper Co.)	Espanola, Ontario
The Ontario-Minnesota P&P Co. (formerly Kenora Paper Mills, Ltd.)	Kenora, Ontario
Ontario Paper Co. (owned by Chicago Tribune)	Thorold, Ontario
Pacific Mills, Ltd.	Ocean Falls, British Columbia
P. C. Powell River Co.	Powell River, British Columbia
Quebec North Shore Paper Co. (owned by Chicago Tribune; mill constructed late 1930s)	Baie Comeau, Quebec
Spruce Falls Power & Paper Co.	Kapuskasing, Ontario

Sources for tables 4a and 4b: Dominion Bureau of Statistics, *The Pulp and Paper Industry in Canada, 1948* (Ottawa, Ontario: King's Printer, 1949), pp. 27-28.

Table 5 New capacity in the southern United States, 1953–59

Year	Company	Location
1953	Bowater Southern Paper Co. opens new mill	Calhoun, Tennessee
1956	Bowater Southern Paper Co. adds third paper machine to mill	Calhoun, Tennessee
1956	International Paper Co. opens new mill	Mobile, Alabama
1956	Southland Paper Mills, Inc., adds third machine to mill	Lufkin, Texas
1958	Bowater Southern Paper Co. adds fourth machine to mill	Calhoun, Tennessee
1958	International Paper Co. opens new mill	Pine Bluff, Arkansas
1959	Southland Paper Mills, Inc., adds fourth machine to mill	Lufkin, Texas
1959	Noralyn Paper Mills starts construction of a newsprint mill using hardwood-based stock	Baton Rouge, Louisiana

new newsprint mills, none became operational until the MacMillan Bloedel plant at Port Alberni opened in September 1957, the only new newsprint mill opened in Canada during the 1950s. Throughout the decade, Canadian newsprint companies expanded production by upgrading their plants and adding new paper machines to take advantage of new papermaking technology.

There was little chance by the end of the decade that new newsprint mills would be built in Canada. A recession from 1956-59 seemed to confirm the wisdom of investor caution. Given the price of newsprint, the potential return on a new mill operating in Canada was not attractive.

Investors were rightly concerned about the position of Canada's newsprint industry. New mills being built in the southern United States used the latest technology, exploited low wood costs, were closer to markets, and had lower labor costs than their Canadian counterparts. In 1955-56 researchers for the Royal Commission on Canada's Economic Prospects noted the presence of this new source of newsprint for the American market. In its *Final Report* the commission confirmed that paper from these new mills affected traditional Canadian markets for newsprint and forecast that the trend would continue. During the 1950s two new mills opened in the southern United States, while other operations expanded (see table 5). As a result, domestically produced newsprint consumed in the United States rose after 1953; by 1959 it had reached 27 percent (see figure 14).

Canadian industry experts explained these new developments by reasoning that temporary tax advantages and low wood costs created competitive advantages for southern newsprint over Canadian newsprint. They predicted that in the 1960s wood costs for southern mills would rise to equal or exceed those of Canadian mills and that the tax advantages would expire. They also postulated that since the advanced technology southern mills used was also available to Canadian mills, Canadian mills would soon adopt these methods. This included technology to make acceptable newsprint from hardwood species. As a result they expected Canada to regain its monopoly positions on supply of newsprint to North America and to recapture overseas markets as well. The accuracy of these predictions was soon tested.

Figure 13 **Canadian newsprint production vs. percentage of capacity, 1945–59**

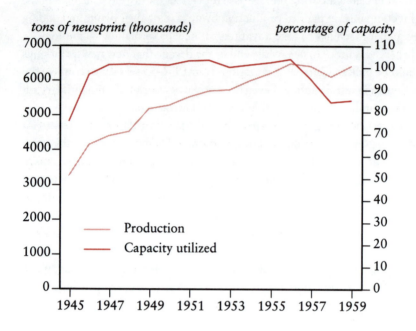

As newsprint production grew, Canadian paper manufacturers declared shareholder dividends that doubled from 2.5 percent of operating revenues in 1945 to 5 percent in 1948. The Newsprint Association of Canada noted these dividends were the first that most companies paid in twenty years or more. Steady growth lasted from 1945-56, and Canadian companies maintained production at full capacity.

Figure 14   **Newsprint manufactured in the United States, 1945–59**

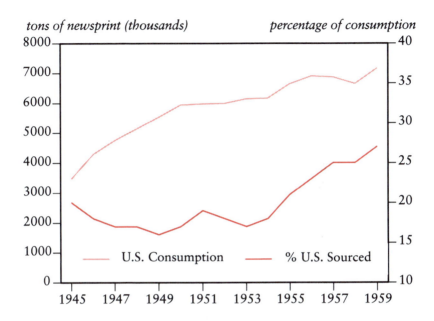

During the 1950s two new mills opened in the southern United States while ongoing operations expanded. As a result, domestically produced newsprint consumed in the United States rose after 1953.

Modern Times

Canada's economy grew steadily from 1961-69 following the recession that marked the end of postwar expansion. Years of fluctuating economic activity that resulted in little real growth followed this period.

During the 1980s the Canadian federal government attempted to gain control of the economy. The Bank of Canada manipulated interest rates in an effort to tame inflation, but at the price of an artificially high value for the Canadian dollar. An overvalued dollar negatively affected Canada's ability to produce competitive exports, weakening the nation's ability to compete in the American market with low-cost producers from other parts of the world.

Canada's newsprint manufacturers have ridden the forces influencing the Canadian economy since the 1950s. The industry has also dealt with several new factors. These include increases in the relative price of fuel, a long period of high inflation, public and government pressure for the industry to adopt environmental safeguards, growth in new sources of newsprint supply from outside North America, the need to lower production costs to remain competitive, and the need to adapt to the changing structure of Canadian forests. In addition, the structure of the industry itself has changed. Companies have been taken over, new companies have bought into the industry, and several new mills have recently been constructed. Although newsprint is still the largest single product of Canada's pulp and paper industry, it has decreased in importance over the past thirty years.

The Environment

While Canadian newsprint makers struggled with economic problems besetting the industry, environmental issues heavily influenced their tactics. Environmental controversy has followed Canada's forest industries since the beginning of the conservation movement in the late-nineteenth century. Controversy most often surrounded forest utilization and logging practices, although river pollution caused by dumping effluent from mills has also been an issue.

Following a landmark study in 1988, Swedish pulp and paper mills started to reduce their use of elemental chlorine. Elemental chlorine (AOX) and other chlorine-based compounds are used to bleach some papers, but are not usually used in newsprint production. Following the reductions by Swedish mills, Canadian provincial governments experienced considerable pressure to require that all mills stop using chlorine and its associated compounds as a bleach in the pulping process. The implication prompting this demand was that all pulp and paper mills used chlorine and that all chlorine-based compounds were harmful to fish.

Neither point was true. Chlorine-based compounds are used only in a minority of pulping processes (those associated with the production of kraft paper). Canadian research also showed that although something in mill effluent harmed fish stocks, it probably was not the AOX compounds, although some kraft pulp mills released trace amounts of dioxins.

Environmentalists in Europe demanded that Canada follow Sweden's lead and cease using chlorine and its compounds in pulp manufacturing. Some Canadians suspected that environmentalists' demands were partly driven by Swedish manufacturers. Forcing Canada to meet this requirement, while not fully addressing the general water quality problem, might give Swedish and other European mills a competitive advantage at Canadian producers' expense.

Despite lack of substance to the demands, Canada's pulp and paper manufacturers reacted swiftly and decisively. Within a year, they practically eliminated release of dioxins; mills that could do so began using alternatives to chlorine-based bleaches (such as hydrogen peroxide); and Canadian production of totally chlorine free (TCF) pulp increased. Totally chlorine free production went beyond the simple chlorine free product from Swedish mills. Unlike chlorine free, which may contain chlorine dioxide, totally chlorine free production uses no chlorine compounds at the mill. Canada currently leads the world in totally chlorine free pulp production, and exports of this product to Europe are increasing. Industry is funding research aimed at discovering the chemical compounds that cause problems in the effluent and eliminating them from the production process. Eliminating chemicals from the process is expected to be more cost effective than constructing relatively expensive secondary treatment plants.

Table 6   Canadian wood products as percentage of total, 1948, 1960, 1985

Product	1948	1960	1985
Wood pulp	20%	22%	21%
Newsprint	63%	58%	35%
Paperboard	7%	9%	8%
Building boards	2%	2%	1%
Papers: fine, book, tissue, wrapping, and other products	7%	8%	8%
Converted to products made of or containing paper	1%	1%	27%

Sources: Dominion Bureau of Statistics, *The Pulp and Paper Industry in Canada, 1948* (Ottawa, Ontario: King's Printer, 1948); Stefan Willie, *The Pulp, Paper, and Allied Industries in Canada* (Oakville, Ontario: Aktrain Research Institute, 1988).

This controversy did not, however, address problems associated with effluent quality. Until 1992, Canadian pulp and paper mills released much of their liquid waste into the environment after primary treatment. This waste contained chemicals harmful to aquatic life although the precise nature of these compounds had not been determined. In May 1992, the federal government promulgated new effluent regulations under the Fisheries Act. These regulations set new standards for all mill effluent, including biological oxygen demand (BOD), suspended solids, and acute lethality. Canadian mills have until the end of 1995 to construct and install secondary treatment plants to insure that effluent conforms to these new standards.

The industry is also investing heavily in the use of recycled paper. Recent government regulations in the U.S. have required that newsprint contain a minimum level of recycled material. The amount of recycled material required varies by state, from 25 percent to 50 percent. Driving forces behind these regulations are the reduced capacity of landfill sites and the desire to reduce the amount of waste discarded annually. Recycled newsprint, however, has problems. Some commentators argue that it is too expensive to collect and process so its use will soon cease. The basis for this argument is that collection is carried out at the municipal level and subsidized by local governments. Subsidies acceptable during periods of economic expansion become unpopular during economic slowdowns. To some this means that the use of recycled paper is ephemeral and investment in expensive de-inking plants is a mistake. Others argue that paying the cost of recycling and thereby reducing the demand for new raw material is a cost of maintaining a high standard of living. Regardless, several paper recycling plants have recently been constructed across Canada to utilize waste paper and to meet recycled content requirements.

These issues typify the current state of environmental issues facing the Canadian newsprint industry. Groups concerned about environmental issues are no longer regionally based, and they use international contacts to communicate viewpoints to a wide audience. It may be instructive for Canadian pulp and paper manufacturers to study why Canada lost the battle over fur-trapping and the annual fur-seal hunt. The fight was lost when a coalition of North American and European groups destroyed the European market for these products.

Changes in the Structure of the Pulp and Paper Industry

Since the end of the Second World War, the scope of Canada's pulp and paper industry has broadened and diversified. Although the classification of different products varies from source to source, there is some overlapping. Table 6 shows the growth in the variety of final products the industry directly produces. Items made of paper include a wide range of consumer products from asphalt roofing, boxes, bags, and wallpaper to writing pads, envelopes, diapers, towels, and paper cups. Prior to 1960 many of these products were made in Canada by secondary manufacturers, although most were imported from the United States. As a result of vertical integration, the proportion of wood pulp and newsprint that is an end product in the manufacturing process has fallen significantly. This change has led one commentator to observe that Canada's paper manufacturers are successfully seeking and capturing product niches for themselves. The process enhances the long-term stability of the industry and creates opportunities for future growth.

Figure 15   U.S. newsprint consumption, 1960–90

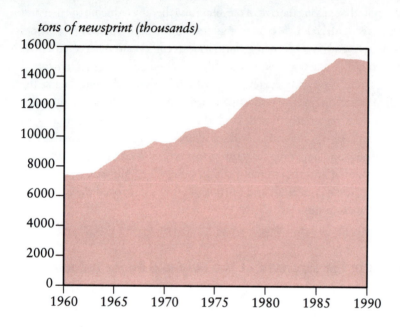

One factor in the overall growth of newsprint consumption has been the expansion of advertising flyers delivered door-to-door or included in newspapers. This use of newsprint has meant that consumption has shown remarkable resistance to fluctuations in the North American economy even while general newspaper readership has declined.

An Economic Overview of Newsprint

During the past three decades American gross consumption of newsprint has grown steadily (see figure 15). Per capita consumption, however, has hardly changed over the past thirty years, maintaining a level between twenty and twenty-five pounds per person. One factor in the overall growth of newsprint consumption has been the expansion of advertising flyers delivered door-to-door or included in newspapers. This use of newsprint has meant that consumption has shown remarkable resistance to fluctuations in the North American economy, even while general newspaper readership has declined.

In Canada newsprint production has increased and growth has been steady (see figure 16). Viewed year to year, production has been heavily influenced by the general state of the North American economy. This is particularly noticeable during recessions that occurred at the end of the 1950s, the mid-1970s, the early 1980s, and 1989 to the present.

Despite fluctuations, Canadian newsprint manufacturers have maintained production at between 78 and 100 percent of capacity. As in all the post-Second World War decades, this level has been achieved by upgrading existing mills, replacing older equipment, or simply adding new machinery. During the 1980s federal grants to the industry assisted in the upgrading process. Commentators now consider this federal assistance ill advised. Besides reinforcing the basic conservatism of Canadian newsprint manufacturers, it delayed the inevitable date when older machines had to be replaced. Also, the grants were available when profits from newsprint production were high. This meant that a sixty-year-old machine, with minor improvements, could continue profitable operation. No new mills became operational until the 1990s. During this decade, several companies that formerly produced only pulp began producing newsprint and some new newsprint companies started operation (see table 7).

Factors Influencing the Price of Newsprint

The price of newsprint responds to world economic events. During the 1960s the price of newsprint f.o.b. New York was remarkably stable, remaining at $134 per ton for 32 lb. paper until 1966 (see figure 17). From 1966–73 prices rose slowly to an average $175 per ton. In 1973 oil prices started to rise, and in North America the sudden price increases contributed to a downturn in the economy and boosted inflation. Operating costs for the pulp and paper industry increased, since fuel costs are second only to the cost of fiber for

Figure 16   Canadian newsprint production vs. percentage of capacity, 1960–90

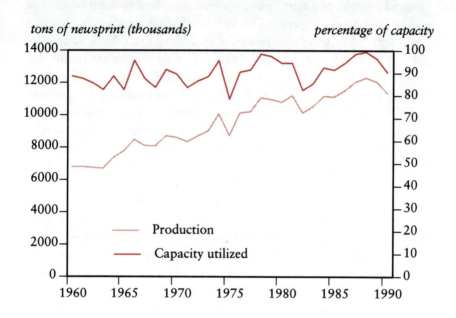

Despite fluctuations, Canadian newsprint manufacturers have maintained production at between 78 and 100 percent of capacity. As in all the post-Second World War decades, this has been achieved by upgrading existing mills, replacing older equipment, or simply by adding new machinery.

Figure 17   Newsprint price, f.o.b. New York, 1960–90

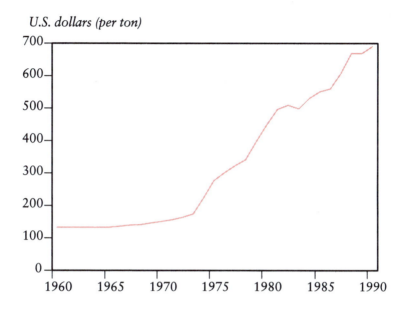

A sudden increase in oil prices in the early 1970s contributed to a downturn in the economy and boosted inflation throughout North America. Operating costs for the pulp and paper industry sharply increased, driving up the price of newsprint.

Table 7   Canadian newsprint manufacturers, June 1991

Company	Mill location	Annual capacity (tons)
Abitibi-Price, Inc. (founded in 1974 with the takeover of Price Brothers & Co. by Abitibi Power & Paper Co.)	Alma, Quebec Beaupré, Quebec Thunder Bay, Ontario Grand Falls, Newfoundland Iroquois Falls, Ontario Kenogami, Quebec Pine Falls, Manitoba Stephenville, Newfoundland Thunder Bay, Ontario	127 10 139 177 252 33 167 169 157
Alberta Newsprint Co. (new; uses poplar resources)	Whitecourt, Alberta	194
Atlantic Newsprint Co. (new; recycles newsprint)	Whitby, Ontario	85
Boise Cascade Canada, Ltd. (formerly Ontario-Minnesota P&P Co.)	Kenora, Ontario	268
Bowater Mersey Paper Co. (formerly Mersey Paper Co.)	Liverpool, Nova Scotia	218
Canadian Pacific Forest Products, Ltd. (merger of Canadian International Paper Co. and Great Lakes Forest Products)	Dalhousie, New Brunswick Gatineau, Quebec Thunder Bay, Ontario Trois Rivières, Quebec	301 481 387 270
Daishowa Forest Products, Ltd. (formerly Anglo-Canadian P&P Mills, Ltd.)	Quebec, Quebec	429
Domtar, Inc. (formed in 1961 by purchases from Howard Smith Paper and Lake St. John Power & Paper Co.)	Dolbeau, Quebec Donnacona, Quebec	150 35
Donohue, Inc.; Donohue Malbaie, Inc.; & Donohue Normick, Inc. (originally part of Abitibi Power & Paper Co.)	Clermont, Quebec Clermont, Quebec Amos, Quebec	153 177 172
Finlay Forest Industries, Inc.	MacKenzie, British Columbia	167
Fletcher Challenge Canada, Ltd.	Crofton, British Columbia Elk Falls, British Columbia	463 442

Table 7 (continued)

Company	Mill location	Annual capacity (tons)
Gaspesia Pulp & Paper (originally part of Price Brothers & Co.)	Chandler, Quebec	234
Gold River Newsprint, Ltd.	Gold River, British Columbia	204
Howe Sound Pulp & Paper Co.	Port Mellon, British Columbia	95
J. D. Irving, Ltd. (originally produced pulp & kraft paper)	St. John, New Brunswick	334
James Maclaren Industries, Inc.	Masson, Quebec	191
Kruger, Inc. (formerly Brompton P&P Co. and Kruger P&P Co.)	Bromptonville, Quebec Corner Brook, Newfoundland Trois Rivières, Quebec	215 322 296
MacMillan Bloedel, Ltd.	Port Alberni, British Columbia Powell River, British Columbia	140 450
Quebec & Ontario Paper Co. (still owned by the Chicago Tribune; shares offered to the public 1992)	Baie Comeau, Quebec Thorold, Ontario	458 313
F. F. Soucy, Inc.	Rivière-du-Loup, Quebec	160
Spruce Falls Power & Paper Co., Ltd.	Kapuskasing, Ontario	314
Stone-Consolidated, Inc. (succeeded Consolidated Bathust)	Grand-Mère, Quebec Port Alfred, Quebec Shawinigan Falls, Quebec Trois Rivières, Quebec	16 378 330 110
Stora Forest Industries, Inc. (new mill built by Swedish interest; uses wood from Cape Breton, Nova Scotia)	Point Turner, Nova Scotia	174

Sources: *Pulp and Paper Annual and Directory* (Toronto, Ontario: Southam Business Publications, 1991); Canadian Forest Service, *Inter-corporate Ownership in the Canadian Pulp, Paper and Paperboard Industry* (Ottawa, Ontario: Queen's Printer, 1988); Barrie McKenna, "How a Megaproject Became a Millstone," *Toronto Globe and Mail*, 22 December 1992, p. B18; Kimberly Noble, "Forest Giants Lose Ground," *Toronto Globe and Mail*, 3 November 1992, p. B6; Kimberly Noble, "Publishers Balking at Proposed Hikes," *Toronto Globe and Mail*, 12 August 1992, p. B1; Robert Williamson, "Pulp Cleanup May Be a Waste of Money," *Toronto Globe and Mail*, 23 December 1992, p. A1.

Canadian paper producers. Fuel costs are also hidden in fiber costs, especially for companies that truck logs long distances. Not surprisingly, the price of paper rose from $175 per ton to $400 by 1980.

In 1974 the industry agreed to switch from making paper that weighed 32 lbs. per 500 sheets, to 30 lbs. per 500 sheets. This represented a 6.25 percent increase in the number of sheets of newsprint that could be cut from a roll of paper, lowering the cost to consumers. Experience during the Second World War showed that using this paper weight maintained visual quality while preventing ink bleed-through. But switching to a lighter grade of newsprint itself did little to slow the rise in price. Comparing the index of the price of newsprint with the general manufacturing index, however, reveals that for much of this period changes in the price of the product were closely tied to changes in the general index. In the second half of the 1980s, this correlation changed and the index of the contract price of Canadian newsprint exceeded that of the general index.

Market Changes

The most significant economic feature of the industry during the years since 1960 is the degree to which the importance of Canadian newsprint to the American market has decreased (see figure 18). In 1960 Canada exported approximately 78 percent of newsprint production to the United States. Since that time, this ratio has dropped to the current level close to 70 percent. The significance of these figures is apparent from the American point of view: in 1960 over 70 percent of newsprint consumed in the United States was manufactured in Canada; currently this figure is close to 50 percent. In the intervening decades the proportion of domestically produced newsprint used in the United States has increased from 27 percent to 48 percent. This means that the United States will soon gain control of its newsprint supply, a situation that has not existed since the mid-1920s.

Historically, Canada has not needed the American market as much as America needed Canadian supplies. But as we have seen, Canadian newsprint sales to the U.S. are not expanding as fast as the American market, so the importance to the United States of Canadian production is decreasing while the importance to Canada of the American market is increasing. This means Canada is not competing favorably on the world market.

Figure 18   U.S. imports of Canadian newsprint, 1960-90

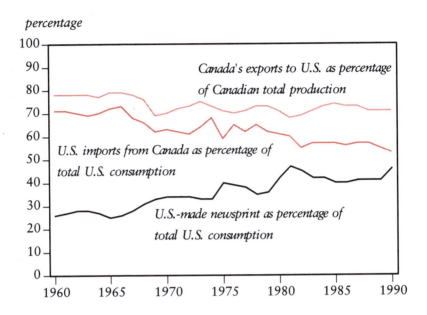

Canadian newsprint sales to the U.S. are not expanding as fast as the American market, and U.S. domestic production is simultaneously increasing. For example, as the above graph shows, in 1960 Canada exported approximately 78 percent of its total newsprint production to the United States; current Canadian exports to the United States have fallen to about 70 percent of total production. Canada manufactured over 70 percent of newsprint consumed in the United States in 1960; this figure is now close to 50 percent. The United States produced about 27 percent of the newsprint consumed in the United States in 1960; today the United States produces almost 50 percent of the newsprint the country consumes.

Canadian Newsprint Manufacturers

Table 7 shows changes in the organization of several companies. These include the formation of Domtar from the St. Lawrence Corporation and the Howard Smith Paper Company in 1961, while Canadian Pacific Railways formed Canadian Pacific Forest Products by purchasing Canadian International Paper and Great Lakes Forest Products in the 1980s. In 1989 Stone Container Corporation purchased Consolidated Bathurst from Power Corporation, a purchase worth $2.6 billion that increased Stone's debt to capitalization ratio from 40 percent to 70 percent. The major event of the 1970s, however, was Abitibi Paper Company's purchase of Price Co. in November 1974. The reasons for these investment decisions shed light on industry leaders' attitudes toward the industry because they present evidence that managers focus on maintaining high earnings.

In the early 1970s Abitibi Paper actively sought to expand production, preferably by increasing its share of the American paper market. The company had three options to achieve this objective. The first tactic was to build one or more new mills. This was ruled out because of the high interest rate on money borrowed to finance construction. Abitibi was receiving a 15 percent (after tax) return on its existing mills and research showed that in the short term a new mill would return 8 percent to 10 percent on the $200 million invested.

The second approach was to purchase an existing company in the United States. The company rejected this because it feared such a move might violate U.S. federal antitrust laws. Even if the purchase were lawful, such a course might result in Abitibi losing its status as a Canadian company, making it difficult in the future for the company to purchase other Canadian companies.

The third tactic was to take over an existing Canadian operation that had modern mills and a developed niche in the American market, which would help keep earnings at then current levels. Price Co. closely fit this criteria for a takeover bid because it showed promise of boosting Abitibi's profitability. Price had updated existing mills, low debt load, high retained earnings, and undervalued shares. The shares of many companies, including Abitibi, were undervalued at the time because of concern over an impending recession and the prospect of further increases in inflation due to the OPEC oil maneuverings. As a result of the Price takeover, Abitibi nearly doubled production capacity for about half the cost of building new mills. Abitibi increased profitability because Price mills had profits near the same level as Abitibi, close to 15 percent.

This concern with short-term earnings may have resulted from Abitibi's attempts to redress its failure to pay dividends during the 1930s and 1940s, or it may have been that Canadian tax laws overly restricted industry's ability to recoup investments. In any case, the takeover did not alter the new company's need ultimately to upgrade its plant in order to maintain competitiveness.

Canadian International Competitiveness

Canada's federal forestry department has shown that for most of the 1980s exchange rates, rather than lowered production costs, were the major factor determining Canada's competitive position. Typically newsprint production costs are lowest in the southern and western United States. When the Canadian dollar is appreciably lower in value than its American counterpart, Canada can produce newsprint at lower cost than Sweden or Finland, but still not lower than American domestic producers. For example, from 1984-88 Canadian production and delivery costs rose 7 percent while Finnish costs fell 2 percent. Yet Canada maintained its position largely because the value of the Canadian dollar fell enough to compensate for the rise in production costs. Later in the decade, the Canadian dollar rose and Canada dropped below Sweden in competitiveness but still remained ahead of Finland. The strength of the Canadian dollar in the years since 1988 (although it has been historically weak—see figure 19) led Canadian newsprint producers to institute pricing policies they adopted in the 1930s, which ultimately severely damaged the industry.

There is, however, a significant difference between the 1930s and the 1990s. During the 1930s companies offered discounts on the contract price of newsprint because of general overproduction. Discounts in the late-1980s were not a response to overproduction but an effort to maintain Canada's competitive position vis-à-vis other producers, especially European producers. In 1988, for example, the contract price of newsprint in the eastern United States was $580 per ton but consumers actually paid between $518 and $545 per ton. An attempt to raise the contract price to $612 per ton in the period March 1989 to June 1990 resulted in even deeper discounts. By the end of 1992, while the contract price remained stable, consumers took advantage of discounts to pay an average $379 per ton for newsprint. In 1991 Canadian newsprint producers lost $250 million as a result of discounting. This deficit contributed to the overall $612 million loss the industry suffered that year.

Figure 19 Value of Canadian dollar in terms of U.S. dollars, 1952–89

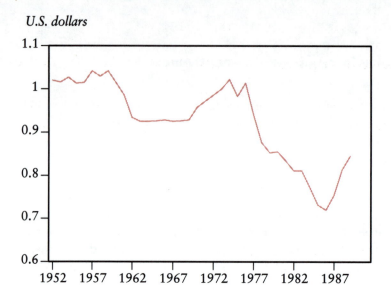

The value of the Canadian dollar in relation to the American dollar has been an important consideration in the competitiveness of the newsprint industry.

Oversupply Not the Problem

Many Canadian newsprint producers blame current economic problems on oversupply, meaning that demand for paper is not high enough for mills to operate profitably. This is the same terminology used in the 1930s, when oversupply *was* the problem. North America's capacity to produce newsprint at that time outstripped demand, and desperate mills flooded the market with paper at cut-rate prices in an effort to generate enough cash to stay afloat. But in today's environment it is unlikely prices would increase even if Canadian production matched or fell short of demand (a situation of undersupply) since consumers have diverse alternative sources. Thus, Canadian oversupply is not the cause of low newsprint prices and the industry's consequent malaise.

The source of problems facing Canada's newsprint industry is that production costs are too high, especially fiber costs. If Canadian mill operating costs were as low as those of Scandinavia, and fiber costs lower than those in the United States, the Canadian newsprint industry would be competitive. But this is not yet the case. Comparison of the production costs in Canada, the United States, Sweden, and Finland shows that Canadian wood costs are higher than those of companies in the southern and western United States but considerably lower than those in the two Scandinavian countries. The Scandinavian countries instead have lower labor costs because of large, modern, efficient papermaking equipment. Canadian labor costs are comparable to those in the United States (many studies show them to be higher, but lower Canadian energy costs account for this differential).

Machine Size a Factor

One factor contributing to the higher cost of producing newsprint in Canada is the comparatively small size of Canadian papermaking machines. Despite assisted investment in upgrading machines and investment in installing new ones, the average Canadian machine has only 57 percent of the capacity of an average Finnish machine and 79 percent of the capacity of an average machine in the United States. Competitors' machines are also newer and have lower operating costs. The relative small size of Canadian papermaking machines is due to Canadian papermakers' conservative attitude toward expansion investment, a shortcoming many commentators have criticized. This conservatism grew out of problems the Canadian newsprint industry faced after overexpansion during the 1920s and 1930s. A tradition evolved that governs new mill expansion and explains why Canadian mill owners tend to "tinker" with their plants rather than replace them.

This traditional method involved building the mill one section at a time, starting with the pulp operation. As each section came into production, it financed construction of the next. This approach was inefficient because it restricted the size of the plant and resulted in the average small size of Canadian papermaking machines. It also made it difficult to build a plant that used the latest technology in an integrated whole because by the time the last machine was installed earlier facilities were often technologically outdated. The most recent new newsprint mills in Canada, however, have been designed and constructed as state-of-the-art units.

The Changing Structure of Canadian Forests

Another major problem facing pulp and paper manufacturers in Canada is that their forestry activities were originally adapted to harvesting existing, naturally regenerating forests. Mills were designed and built to exploit the forest as it existed before operations began. In many other countries, particularly in Europe, this problem is circumvented by reliance on artificial plantations. Some industry sources have said that plantation forestry is too expensive for Canada because tree growth is too slow and labor costs are too high for it to be profitable. There are also questions about the ecological viability of planting large areas with one species of tree (monoculture).

Despite these criticisms, the production of tree seedlings in Canada has expanded since the early 1970s. The number of seedlings planted has grown from approximately 100 million per year to 750 million in 1988. At the same time, large-scale programs were started to thin the many young stands of naturally regenerated trees growing too close together and thus reducing growth rates. These programs, considered successful, indicate that large volumes of pulpwood will be produced in the near future at rotations of forty to fifty years.

A New Approach to Fiber Production

North America today has about three times the population it had in 1913. Per capita newsprint consumption, however, has remained stable for several decades. This means demand is increasing at a lesser rate than population growth, so newsprint manufacturing is a mature industry. The challenge to science and industry, then, is to produce newsprint in a manner that fulfills criteria that arise from expanding population and increased demands on natural resources. Land managers and newsprint manufacturers must fill the demand for newsprint while respecting amenity values.

Stewardship demands come from the general values of society and have little to do directly with manufacturing, but the newsprint industry must respect these demands. Smaller management units, intensively administered on an ecologically sustainable basis, are the second prong of the future for Canada's newsprint industry, just as they are for forestry in general.

The major difficulty has been in overcoming attitudes. The general public, as well as investors, needs information about how forests grow and respond to management. They also need to know about new approaches to paper production, especially about how such methods work well in other countries and are working well as recently applied in Canada. This is true in the context of two important history lessons: (1) A viable newsprint manufacturing industry in Canada is possible and (2) events of the 1920s and 1930s cannot be permitted to continue to influence the future of the industry.

These admonitions are clearly reflected in changes that have occurred within the forestry sector during the last decade. A national forest strategy was developed through extensive public consultations over a two-year period, culminating in a National Forest Congress in 1992 and described in *Sustainable Forests: A Canadian Commitment* (Canadian Council of Forest Ministers, 1992). The strategy described the evolution of forestry practices, presented statements of values and vision, and outlined commitments to which industry, government, labor, and conservation groups subscribed. The commitments are designed to continue the shift to sustainable forestry based on a stewardship ethic and public participation. Forest industries have also developed codes of practice, and provincial governments are writing forest strategies for their own forests based on these principles.

Forest stewardship and sustainability are the hallmarks of the future for Canada's newsprint industry.

Suggested Reading

Anderson, Hugh. "Well-intentioned Government Scheme Backfires on Pulp-and-paper Industry." *Montreal Gazette*, 13 February 1992.
A Source of Pride: Canada's History in Pulp and Paper. Toronto, Ontario: Maclean-Hunter, 1992.
Cagampan-Stoute, Caroline. "Top Performers." *Pulp and Paper Journal* 45 (June/July 1992): 37-42.
Canadian Council of Forest Ministers. *Sustainable Forests: A Canadian Commitment*. Hull, Quebec: Canadian Council of Forest Ministers, 1992.
Canadian Forest Service. *Inter-corporate Ownership in the Canadian Pulp, Paper and Paperboard Industry*. Ottawa, Ontario: Queen's Printer, 1988.
Carmichael, Herbert. "Pioneer Days in Pulp and Paper." *The British Columbia Historical Quarterly* 9 (July 1945): 201-12.
Carruthers, George. *Paper in the Making*. Toronto, Ontario: Garden City, 1947.
Cooke, Maxine, ed. *Pulp and Paper Directory of Canada-1970*. Gardenvale, Quebec: National Business Publications, 1971.
Dagenais, Marcel G. "The Determination of Newsprint Prices." *Canadian Journal of Economics* 9 (August 1976).
Dominion Bureau of Statistics. *The Pulp and Paper Industry in Canada, 1948*. Ottawa, Ontario: King's Printer, 1948; 1949.
Ellis, Lewis Ethan. *Newsprint: Producers, Publishers, Political Pressure*. New Brunswick, New Jersey: Rutgers University, 1960.
Foster, Peter. *The Blue-eyed Sheiks: The Canadian Oil Establishment*. Toronto, Ontario: Collins, 1979.
Guthrie, John A. *The Newsprint Paper Industry*. Cambridge, Massachusetts: Harvard University Press, 1941.
Haviland, W. E., N. S. Takacsy, and E. M. Cage. *Trade Liberalization and the Canadian Pulp and Paper Industry*. Toronto, Ontario: University of Toronto Press, 1968.
Hayter, Roger. "Corporate Strategies and Industrial Change in the Canadian Forest Product Industries." *The Geographical Review* 66 (April 1976): 209-11.
Howlett, Michael. "The Forest Industry on the Prairies: Opportunities and Constraints to Future Development." *Prairie Forum* 14 (Fall 1989): 233-57.
Hull, James P. "Research at Abitibi Power and Paper." *Ontario History* 79 (June 1987): 167-79.
Kellogg, Royal S. *Newsprint Paper in North America*. New York: The Newsprint Service Bureau, 1948.
Marshall, Herbert, Frank Southard Jr., and Kenneth W. Taylor. *Canadian-American Industry: A Study in International Investment*. Toronto, Ontario: McClelland and Stewart, 1976.

Mathias, Philip. *Forced Growth.* Toronto, Ontario: James Lewis and Samuel, 1971.
Mathias, Philip. *Takeover.* Toronto, Ontario: Maclean-Hunter, 1976.
McKenna, Barrie. "How a Megaproject Became a Millstone." *Toronto Globe and Mail,* 22 December 1992, p. B18.
McQueen, Rod. *Leap of Faith.* Toronto, Ontario: Cowan and Company, 1985.
Nautiyal, J. C. and B. K. Singh. "Long-term Productivity and Factor Demand in the Canadian Pulp and Paper Industry." *Canadian Journal of Agricultural Economics* 34 (March 1986): 21-65.
Noble, Kimberly. "Forest Giants Lose Ground." *Toronto Globe and Mail,* 3 November 1992, p. B6.
Noble, Kimberly. "Publishers Balking at Proposed Hikes." *Toronto Globe and Mail,* 12 August 1992, p. B1.
Perry, J. Harvey. *A Fiscal History of Canada: The Postwar Years.* Toronto, Ontario: The Canadian Tax Foundation, 1989.
Pulp and Paper Annual and Directory. Toronto, Ontario: Southam Business Publications, 1981; 1991.
Reich, Nathan. *The Pulp and Paper Industry in Canada.* Toronto, Ontario: Macmillan, 1926.
Roach, Thomas R. "The Pulpwood Trade and the Settlers of New Ontario, 1919-1938." *Journal of Canadian Studies* 22 (Fall 1987): 78-89.
Royal Commission on Canada's Economic Prospects. *Final Report.* Ottawa, Ontario: Queen's Printer, 1957.
Schlesinger, Arthur M., Jr. *The Crisis of the Old Order.* Toronto, Ontario: Heineman, 1957.
Schreuder, Gerald F., ed. *Global Issues and Outlook in Pulp and Paper.* Seattle: University of Washington Press, 1988.
Stephenson, J. Newell and F. A. Price. *National Directory of the Canadian Pulp and Paper Industries: 1960-1961.* Gardenvale, Quebec: National Business Publications, Ltd., 1962.
Traves, Tom. *The State and Enterprise: Canadian Manufacturers and the Federal Government, 1917-1931.* Toronto, Ontario: University of Toronto Press, 1979.
Thompson, John H. with Allen Seager. *Canada 1922-1939: Decades of Discord.* Toronto, Ontario: McClelland and Stewart, 1985.
Wiegman, Carl. *Trees to News.* Toronto, Ontario: McClelland and Stewart, 1953.
Williamson, Robert. "Pulp Cleanup May Be a Waste of Money." *Toronto Globe and Mail,* 23 December 1992, p. A1.
Willie, Stefan. *The Pulp, Paper, and Allied Industries in Canada.* Oakville, Ontario: Aktrain Research Institute, 1988.

The following sources were used for the statistical information cited in the text:

Newspaper and Newsprint Facts at a Glance: 1992-93. New York: Newsprint Information Committee, 1992.

Editor and Publisher (1952-1992) for information on advertising space purchased in U.S. newspapers in sixty-four cities.

The Newsprint Association of Canada, and its successor the Canadian Pulp and Paper Association, publishes an annual summary of production, demand, and prices for Canada and the world entitled *Newsprint Data*. I have used this source extensively. In the Canadian Federal Government, Statistics Canada and its predecessors also publish extensive annual reviews under a variety of titles. Recently, Forestry Canada through the Forest Sector Advisory Council undertook studies and published the results in *Newsprint Cost Study: 1983-1987*.